Guidance to Cleaning Validation in Diagnostics

Samar K Kundu

ISBN-10: 1499692919
ISBN-13: 978-1499692914

1

Guidance to Cleaning Validation in Diagnostics

Table of Contents

Guidance to Cleaning Validation in Diagnostics

Guidance to Cleaning Validation in Diagnostics

1. Background

This book on Cleaning Validation is intended to address special considerations and issues pertaining to validation of cleaning procedures for equipment used in the manufacture of in-vitro diagnostic products and reagents. This guidance has been prepared only to assist companies in the formulation of cleaning validation programs and provides guidance of developing robust systems.

FDA Guidance to Cleaning Validation of Equipment States the following:

a. *FDA expects firms to have written procedures (SOP's) detailing the cleaning processes used for various pieces of equipment.*

b. *FDA expects firms to have written general procedures on how cleaning processes will be validated.*

c. *FDA expects the general validation procedures to address who is responsible for performing and approving the validation study, the acceptance criteria, and when revalidation will be required.*

d. *FDA expects firms to prepare specific written validation protocols in advance for the studies to be performed on each manufacturing system or piece of equipment which should address such issues as sampling procedures, and analytical methods to be used including the sensitivity of those methods.*

e. *FDA expects firms to conduct the validation studies in accordance with the protocols and to document the results of studies.*

f. *FDA expects a final validation report which is approved by management and which states whether or not the cleaning process is valid. The data should support a conclusion that residues have been reduced to an "acceptable level."*

Guidance to Cleaning Validation in Diagnostics

2. Objective

The objective of the cleaning validation is to verify the effectiveness of the cleaning procedure for removal of product residues, degradation products, preservatives, cleaning agents as well as the control of potential microbial contaminants. In addition one need to ensure there is no risk associated with cross-contamination of active ingredients.

2.1.1 Cleaning procedures must strictly follow carefully established and validated methods of execution. This applies equally to the manufacture of active immunological ingredients, and final diagnostic products. In any case, manufacturing processes have to be designed and carried out in a way that contamination is reduced to an acceptable level.

2.1.2 Cleaning Validation is documented evidence that an approved cleaning procedure is used for the manufacture of active immunological ingredients and final diagnostics products.

2.1.3 Validated Cleaning procedures are necessary for the following reasons:

   a. It is a customer requirement to ensure the safety and efficacy of the product.

   b. It is a regulatory requirement to ensure the quality of manufactured product.

   c. It also assures from an internal quality control and compliance of the process

Guidance to Cleaning Validation in Diagnostics

3. Scope

A Guideline to assure a documented cleaning plan is in place. Six specific areas are addressed in this Guidance document, namely:

3.1 Acceptance Criteria. In diagnostics, it is the critical carryover concentration of active component(s)

3.2 Levels of Cleaning (Closed common equipment for multi products, single use) Bracketing and Worst Case Scenarios (protein, enzymes, microparticles used in immunoassays, microbial agents, chemical additives, detergents, lubricants)

3.3 Determination of the amount of residue (Inside common equipment, contact Surfaces, clean room)

3.4 Cleaning Procedure Validation

3.5 Verification and Validation

3.6 Compliance Plan

4. Definitions Used in Cleaning Validation

The following definitions are commonly used in cleaning validation in in-vitro diagnostics

4.1 Bio-reactive: A material that is the active principal in the Diagnostics Reagent

4.2 Cleaning Agent: Any material used to remove any unwanted contaminants/ residues

4.3 Cleaning Validation: The establishment of documented evidence which provides high degree of assurance that the cleaning procedure consistently remove contaminants/residues to a predetermined level

4.4 Clean-in-Place (CIP): The process of automated (or semi-automated) cleaning equipment in a manufacturing location

4.5 Clean-Out-of-Place (COP): The process of cleaning equipment of equipment's permanent location, but in an automated or semi-automated process

Guidance to Cleaning Validation in Diagnostics

4.6 Dedicated Equipment: Equipment used to manufacture a single product

4.7 Multi-use Equipment: Equipment that processes various products without any specific order

4.8 Equipment Hold Time Before Cleaning: The maximum time that the equipment can be held prior to cleaning

4.9 Contaminant/Residue: Any material to be removed or remaining on equipment surface. Residues for analysis include product/process residues as well as cleaning agent residues

4.10 Visibly Clean: The appearance of no residue on a cleaned surface as evaluated by eye

4,11 Critical Concentration Limit (CCL): This is also called Residual Clinical Limit (RDL) which is the calculated Acceptance Criteria used to avoid carryover of one product to another

4.12 Characterization Study: This study is used to identify the key inputs and outputs of a method or process, collecting data over the entire operating range, estimating the steady-state at optimal operating conditions and build a model describing the parameter relationships at optimal operating conditions that can be used a validation protocol

4.13 Method Validation: It is a process of establishing the performance characteristics and limitations of a method and the identification of the influences which may change these characteristics and to what extent. Which analytes can it determine in which matrices in the presence of which interferences? Within these conditions what levels of precision and accuracy can be achieved? The process for verifying that a method is fit for purpose; i.e. for use of solving a particular analytical problem

4.14 Validation Protocol: An approved controlled document for a particular study that communicated in detail the proposed activities for study. It identifies what the study is expected to accomplish and methods to be used to achieve that purpose

4.15 Validation Report: A document that summarizes in details the results

Guidance to Cleaning Validation in Diagnostics

of the protocol. It identifies what the study accomplished

4.16 Verification: Confirmation by examination and provision of objective evidence that specified requirements has been fulfilled. Verification activities include examination, analysis, demonstration and testing

4.17 Limit of Detection (LOD): LOD's may also be calculated based on the standard deviation of the response (SD) and the slope of the calibration curve (S) at levels approximating the LOD according to the formula: $LOD = 3.3(SD/S)$. The standard deviation of the response can be determined based on the standard deviation of y-intercepts of regression lines.

4.18 Limit of Quantitation (LOQ): LOQ calculation is also based on the standard deviation of the response (SD) and the slope of the calibration curve (S) according to the formula: $LOQ = 10(SD/S)$. Again, the standard deviation of the response can be determined based on the standard deviation of y-intercepts of regression lines.

4.19 Sensitivity: The sensitivity of a method is the rate of change of the measured esponse with change in the concentration of analyte. For instrumental systems, sensitivity is represented from the fitted calibration curve represented by the slope (b) of the calibration curve ($y = a + bx$) and can be determined by a classical least squares procedure, or experimentally, using samples containing various concentrations of the analyte. Sensitivity may be determined by the analysis of spiked or artificially contaminated samples or standards. Analytical sensitivity of a diagnostic assay can be assessed by quantifying the least amount of analyte that is detectable in the sample at an acceptable co-efficient of variation. This can be done by limiting dilutions of a standard of known concentration of the analyte. Another approach is to use end-point dilution analysis of samples from known positive specimens, to define the penultimate dilution of sample in which the analyte is no longer detectable, or at least, is indistinguishable from the activity of negative sera. When the results for the assay under development are compared with other assay(s) run on the same samples, a relative measure of analytical sensitivity can be estimated.

4.20 Change Control: A formalized system by which representatives of appropriate disciplines evaluate proposed changes that may affect the validated status of a given system, process or product. It assures maintenance of a state of cleaniness, any non-conformances detected by the monitoring program, results from periodic reviews. This needs to be investigated and when necessary, corrective action has to be taken. Corrective actions may result in retraining, revalidation and/or introduction of new cleaning methodologies. Any changes in the cleaning process, equipment or cleaning agents must be performed by a change control document and approved by appropriate personnel.

4.21 Re-validation: Performance of a new validation where a validation previously existed

4.22 Corrective action and preventive action (CAPA) are improvements to an organization's processes taken to eliminate causes of non-conformities or other undesirable situations. CAPA is a concept within good manufacturing practice (GMP). It focuses on the systematic investigation of the root causes of non- conformities in an attempt to prevent their recurrence (for corrective action) or to prevent occurrence (for preventive action). Corrective actions are implemented in response to customer complaints, undesired levels of internal nonconformity, nonconformities identified during an internal audit or adverse or unstable trends in product and process monitoring such as would be identified by statistical process control **(SPC)**. Preventive actions are implemented in response to the identification of potential sources of non-conformity.

To ensure that corrective and preventive actions are effective, the systematic investigation of the root causes of failure is pivotal. CAPA is part of the overall quality management system (QMS).

Guidance to Cleaning Validation in Diagnostics

4.23 Statistical Process Control (SPC): Statistical process control (SPC) is a method of quality control which uses statistical methods. SPC is applied in order to monitor and control a process. Monitoring and controlling the process ensures that it operates at its full potential. SPC can be applied to any process where the "conforming product" (product meeting specifications) output can be measured e.g. Control charts.

5. Validation of Cleaning Processes

a. As a general concept, until the validation of the cleaning procedure has been completed, the product contact equipment should be dedicated.

b. In a multi-product facility, validating the cleaning of a specific piece of equipment to be used for multi-product should be rigorously validated using the worst case products used in that equipment.

c. Equipment cleaning validation may be performed concurrently with actual production steps during process development and clinical manufacturing. Validation programs should be continued through full scale commercial production.

d. It is usually not considered acceptable to test-until-clean. This concept involves cleaning, sampling, and testing with repetition of this sequence until an acceptable critical concentration limit is attained.

e. Products which simulate the physicochemical properties of the substance to be removed may be considered for use instead of the substances themselves, when such substances are either toxic or hazardous.

f. Raw materials sourced from different suppliers may have different physical properties and impurity profiles. When applicable such differences should be considered when designing cleaning procedures, as the materials may behave differently.

g. If automated procedures are utilized (Clean-In-Place: CIP), consideration should be given to monitoring the critical control points and the parameters with appropriate sensors and alarm points to ensure the process is highly controlled.

h. During production of several batches of the same product, cleaning between batches may be reduced. The number of lots of the same product which could be manufactured before a complete/ full cleaning is done should be determined.

i. Validation of cleaning processes should be based on a worst-case scenario, challenge of the cleaning process to show that the challenge soil can be recovered in sufficient quantity or demonstrate log removal to ensure that the cleaning process is indeed removing the soil to the required level, and also to use of reduced cleaning parameters such as overloading of contaminants, over drying of equipment surfaces, minimal concentration of cleaning agents, and/or minimum contact time of detergents.

j. At least three (3) consecutive applications of the cleaning procedure should be performed and shown to be successful in order to prove that he method is validated. Equipment which is similar in design and function may be grouped and a worst case established for validation.

6. Cleaning Materials

The production or purchase of all materials that are used during cleaning and/or sanitization operations must be properly controlled with specifications and MSDS. The composition of detergents should be known to the manufacturer. If such information is not available, alternative

Guidance to Cleaning Validation in Diagnostics

detergents should be selected whose composition can be defined. The commonly used cleaning and sanitization agents are shown Table 1.

Table 1: Commonly Used Cleaning and Sanitizing Agents

Cleaning Agents	Source	Cleaning Agents	Source
Advance	Nilfisk-Advance, Inc. Plymouth, MN 55447	Spek-Tak 10,000	U.N.X. Inc. , Greenville, NC 27835
Alconox	Addivant, Middlebury, CT 06762 or other vendors	Ethyl Alcohol	Fisher Scientific or other manufacturers
Amphyl Spray (Lysol)	National Labs Montvale, NJ 07645	Isopropyl Alcohol	Fisher Scientific or other manufacturers
Betadiene	CVS Pharmacy or other Drug Stores	Mandate	Ecolab, St. Paul, Minnesota, MN 55102
Contrad	Fisher Scientific or other vendors	Principal	Kelley Supply, Inc. Colby, WI 54421
Dupanol X1 (Sodium Dodecyl Sulfate (SDS)	Santa Cruz , Biotechnology, Inc. Santa Cruz, CA 95060 or other vendors	Radiac Wash Release	VWR, Fabyan Parkway, Batavia, IL 60510 or other vendors
Starklean 50	JAMO Miami, FL 33166 (supplier): Custom Building Products, Santa Fe Springs, CA 90670 (manufacturer)	Sodium Hydroxide Solution	Fisher Scientific or other manufacturers
Lustrex	Melrose Chemicals, vendor: Revita Battery, LLC, Spokane, WA 99202	Sodium Hypochlorite Solution (Bleach)	VWR or other vendors
Octoxynol purified	Innotech Products, Minneapolis, Minnesota 55414, USA	7X-Detergent	Fisher Scientific or other vendors
Ultrasil 10	Ecolab, St. Paul, Minnesota, MN 55102	7X-O-Matic Detergent	MP Biomedicals,, Solon, OH 44139

Guidance to Cleaning Validation in Diagnostics
7. Cleaning Validation Strategy

In order to have efficient cleaning methods, the following strategy is required. The cleaning strategy of a diagnostic company should be to minimize the risk of product quality. Cleaning validation is directed towards situations or process steps where the potential for carry-over poses the greatest risk to product quality and patient safety. In some cases, the element of reproducibility is limited, cleaning verification may be performed. Cleaning process applies to both cleaning verification and validation activities as described below:

7.1 Cleaning Verification

Cleaning verification typically involves only one lot trial run to demonstrate the effectiveness to release the equipment for further use. Cleaning verification requires the following:

a. All equipment used to manufacture the product must be evaluated to determine the types of residues to be removed (this includes product, previous product, raw materials. microbial residues, intermediates, anti-foaming agents, cleaning agents, lubricants etc.)

b. Proper selection of cleaning agents

c. The equipment configuration and materials of construction must be evaluated to determine the appropriate cleaning methodology

d. Sampling techniques and sites must be determined

e. Test Method must be validated and appropriate to use

f. Justification for acceptance criteria must be documented

g. Cleaning procedures must be documented which must include appropriate process parameters

h. Personnel carrying out the cleaning procedures must be adequately trained

i. Failure to meet acceptance criteria must be investigated

j. The failure must be reported through a CAPA system

7.2 Cleaning Validation

Initial cleaning validation requires a minimum of three consecutive successful operations demonstrating its effectiveness as well as reproducibility. In addition to cleaning verification described above, cleaning validation requires the following described below:

a. Acceptance criteria must be documented prior to execution of the cleaning process

b. Validation of cleaning procedures must reflect normal equipment usage pattern

c. Personnel carrying out the cleaning validations procedures must be adequately trained

d. Failure to meet acceptance criteria must be investigated

e. The failure must be reported through a CAPA system

f. Validated cleaning procedures must be monitored at appropriate intervals using analytical and visual testing to ensure that cleaning procedures remain effective.

Written justification should be provide for the monitoring program as the test method and number of samples that is different that used in cleaning validation

8. Phase 1- Development

The test method development team evaluates the existing analytical test methods that support specific applications. The development team also develops new analytical test methods such as Total Organic Carbon (TOC), Protein assay (BCA), High Performance Liquid Chromatography (HPLC), Conductivity, Phosphate, Chlorine and others for general use to support cleaning for diagnostic products.

Guidance to Cleaning Validation in Diagnostics

8.1 Cleaning Methods Availability

The first thing is to conduct characterization studies for cleaning methods that has the capability of effectively remove contaminants of choice. The characterization studies performed will provide data for assessing which product and cleaning combination results in a worst-case cleaning scenario. In addition, the critical concentration limits (CCL) and ranges for cleaning will be established during these characterization studies. (See the Flow Chart 1 for this development).

Cleaning Validation Strategy

Flow Chart 1: Development Phase

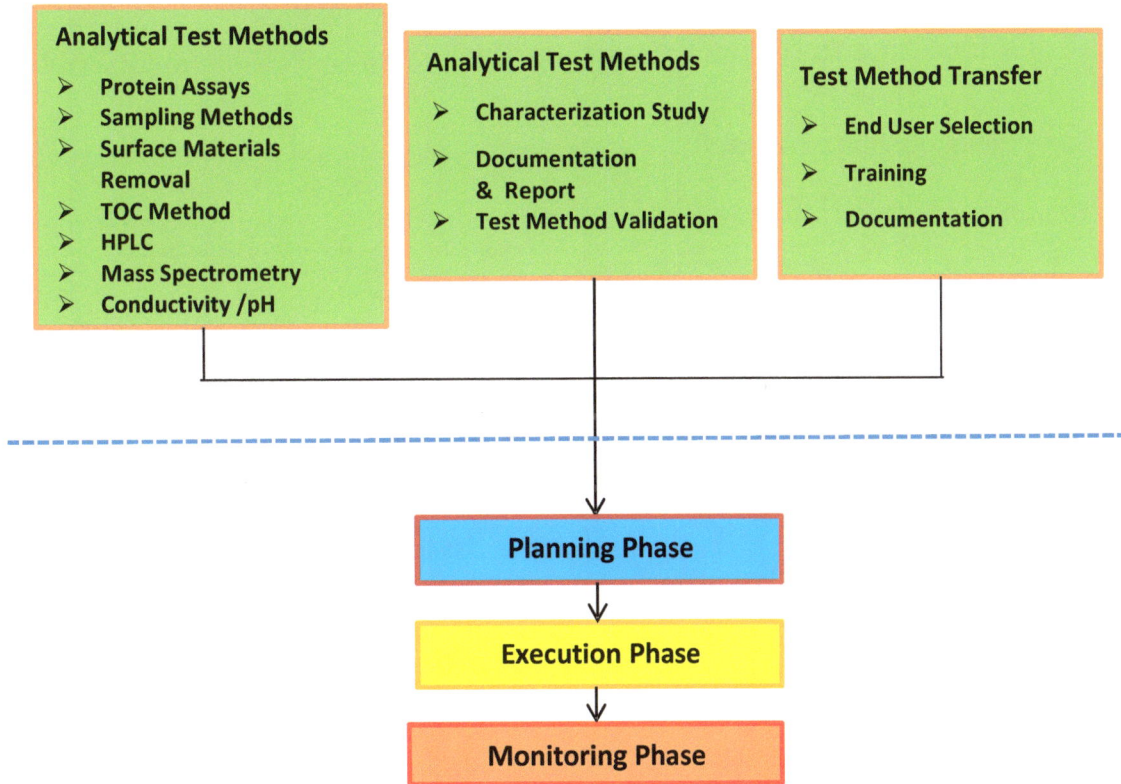

Guidance to Cleaning Validation in Diagnostics

8.2   Analytical Methods to Support Cleaning Validation

The technical test method development team as described in the above flow chart provides support for analytical method development, characterization and validation. The laboratory associated with this team will execute validated test methods. Table 2 describes the list of test methods and sampling techniques that are being used in a diagnostic manufacturing facility.

Table 2: List of Test Methods and Sampling Techniques

Name	Description
Swab Sampling	Technique for Direct Surface Sampling of Equipment
Rinse Sampling	Technique for Indirect Surface Sampling of Equipment
Total Organic Carbon (TOC)	Indicator of the Organic Residue(s) for Swab and Rinse Recovery
Bicinchoninic Acid (BCA) Protein Assay	Quantitation of Residual Protein for Swab and Recovery
Conductivity	Indicator of Ionic Residues
pH	Indicator of Ionic Residues
Anionic Detergent Test	Indicator of Ionic Residues e.g. Sodium docadecyl sulfate (SDS)
Peroxide Test	Indicator for the Presence of Peroxide in detergents, e.g. Tween-20
Microbiological Test	Determination for the Presence of Microbes
Chlorine Test	Indicator for the Presence of Chlorine, e.g., Bleach
Phosphate Test	Indicator for the Presence of Phosphate Residues, e.g. Phosphoric acid
Immunoassay for Test Reagents	Determination of Specific Analytes for Swab and Rinse Recovery

Guidance to Cleaning Validation in Diagnostics

8.3 Method Validation Criteria

For method validation the criteria need to be defined and documented.

These are the major points to consider:

a. Purpose of measurement (why?)

b. Any sample matrices involved?

c. Any interference expected?

d. Expected concentration ranges or levels

e. Any specific environment conditions?

f. Type of instrument and/or equipment for assay performance

g. Type of cleaning equipment (automatic, manual)

h. Construction of the cleaning equipment (glass, stainless steel,. polypropylene, polycarbonate, polyethylene)

i. Reagents and chemicals with defined composition and purity

j. Methods used for preparation of calibrators and reagents

k. Calibrators (standards) and Controls (low, medium and high within the range of calibration curve

l. Use of calibrated instrument(s) and accessories (e.g., pH meter, pipets, balance etc.)

m. Chemicals and reagents with defined purity and composition

n. Stability of prepared reagents for the assay

o. Sampling plan (replicates)

p. Written assay procedure

q. Any regulatory requirements?

8.4 Method Suitability

The components used to manufacture the products need to be reviewed. In cases, where multiple products are made, the review must include all of the products, components and bioactive materials. The worst-case components must be selected from the component listing or soiling compounds. The rationale for this selection will be documented in the validation protocol for the cleaning method. Spiking studies of the worst-case into the easiest to adulterate product will be employed as a tool to establish the critical concentration limit (CCL) as an acceptance limit.

9. Phase 2- Planning

In the planning phase, the cleaning team with the input from manufacturing and/or operations, initiates activities to produce cleaning method validation protocols. The protocols will challenge cleaning methods on equipment to ensure:

a. Cleaning processes that cleans equipment is cleaned below the established acceptance limit.

b. Documents evidence that the finalized cleaning cycles remove residue from the equipment surface and eliminate product carryover and/or cross contamination.

The cleaning team working from Master Equipment List, Master Product List, and Master Cleaning Method List evaluates the following:

9.1 Sampling Plan

Sampling location and sampling schemes based on equipment type and surface area need to be determined and validated with the available validated test methods. If new test methods and sampling techniques are required, these method need to developed and validated by the cleaning test method team.

9.2 Analyte Selection

Each product/equipment combination needs to be evaluated to determine which analytes are most relevant to the cleaning validation. Analytes may include active ingredients used to manufacture the products or the cleaning agents found in the cleaning method. Characterization studies determine which analytes are challenging to the cleaning method. Testing of these analytes utilizes appropriate sampling techniques and test methods.

9.3 Cleaning Method Evaluation

Cleaning methods as being used need to be evaluated for use in equipment and any revisions necessary completed. The cleaning team

Guidance to Cleaning Validation in Diagnostics

based on these characterization studies and evaluations, will write an approved validation protocol to be executed by trained personnel (see Flow Chart 2 for this phase).

Flow Chart 2. Cleaning Validation Team

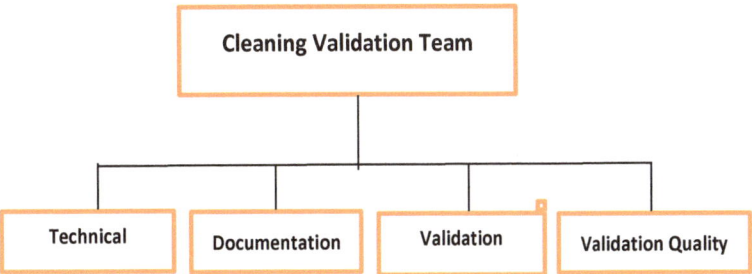

Guidance to Cleaning Validation in Diagnostics

Cleaning Validation Strategy

Flow Chart 3: Planning Phase

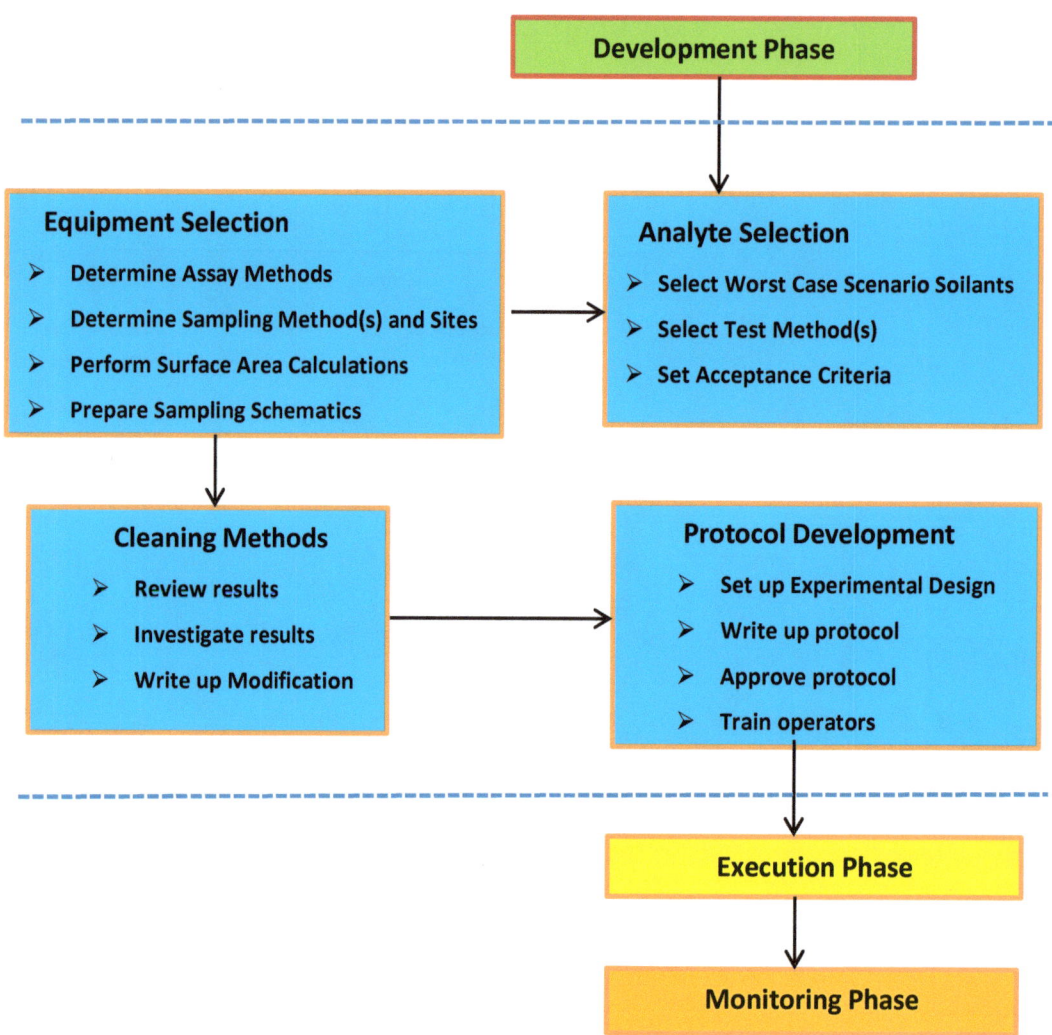

10. Phase 3- Execution

When the validation protocol is approved, the cleaning Validation Team (see Flow Chart 2), and appropriate members from manufacturing and/or operations, performs the following:

Guidance to Cleaning Validation in Diagnostics

10.1 Cleaning Validation Protocol Execution

a. Clean the equipment using appropriate cleaning procedure.

b. Challenge the procedure with worst case soilant(s).

c. Execute the procedure with adequately trained personnel.

10.2 Cleaning Validation Protocol Evaluation

a. Compare the results from the cleaning performance with the acceptable criteria.

b. Investigate any non-conformance(s).

c. Initiate corrective action if necessary.

10.3 Cleaning Validation Report

Write a final validation report and approve by appropriate members (Cleaning Validation, Quality, Operations and/or Manufacturing organization). (See Flow Chart 4)

Guidance to Cleaning Validation in Diagnostics

Cleaning Validation Strategy

Flow Chart 4: Execution Phase

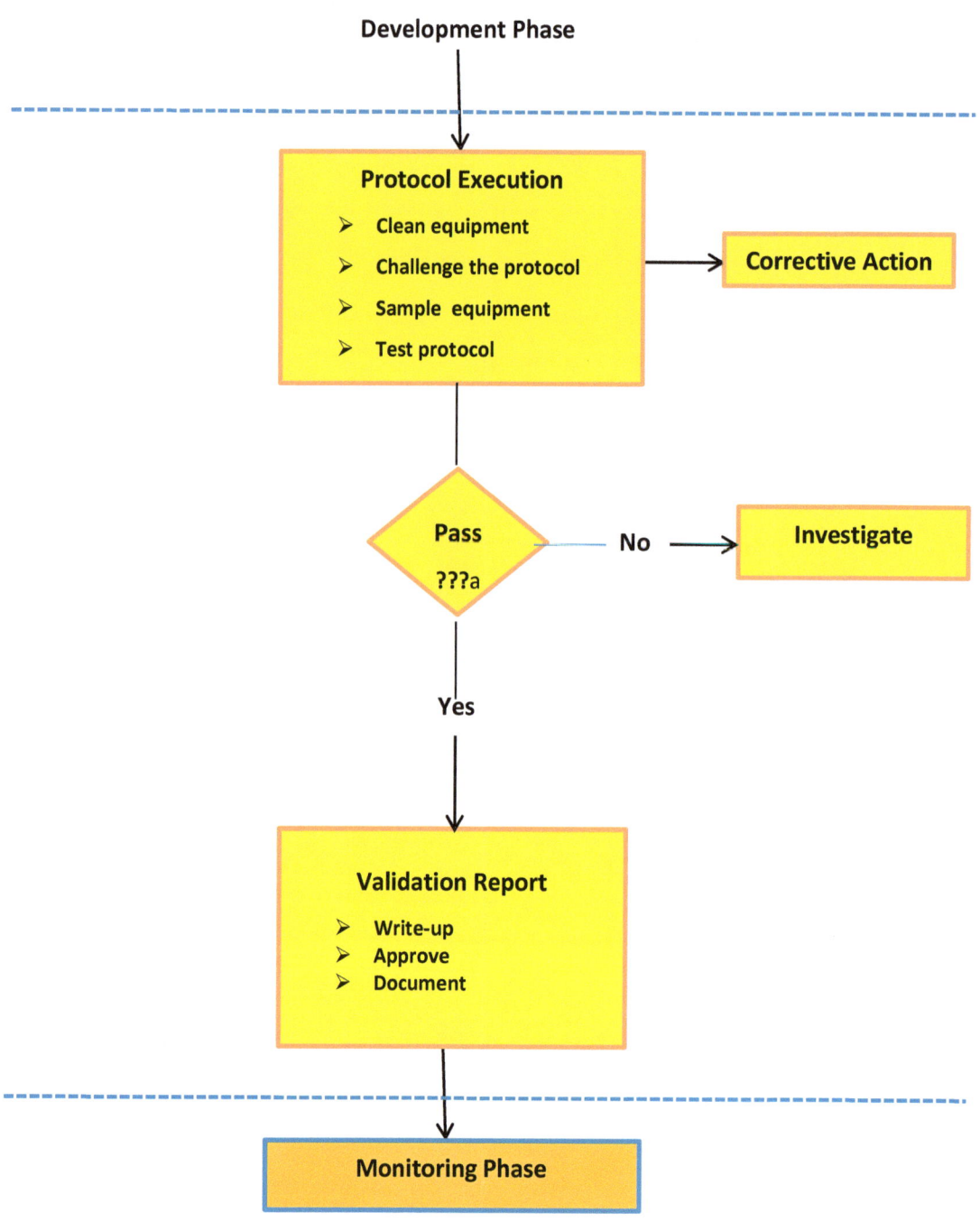

Development Phase

Protocol Execution
- Clean equipment
- Challenge the protocol
- Sample equipment
- Test protocol

Corrective Action

Pass ???a

No → **Investigate**

Yes

Validation Report
- Write-up
- Approve
- Document

Monitoring Phase

Guidance to Cleaning Validation in Diagnostics

11. Phase 4 - Monitoring

Upon completion of the Phase 3 (Execution Phase), the Cleaning Validation Team (CVT) establishes a monitoring program to monitor and maintain the controlled state of cleanliness. The CVT with members of operations and manufacturing develops a monitoring program. This includes:

    a. Equipment to be monitored

    b. Frequency of testing, scheduled and random periods

    c. Establishment of alert values in conjunction with validated acceptance criteria

The monitoring program is an integral component of diagnostic manufacturing operations to ensure the levels of potential contaminants are at or below the acceptable limits. The monitoring program also should include Environmental Monitoring issues to ensure adequate cleaning. The results of Monitoring Program need to be documented. As a result of the documentation if any issue is developed, revalidation of the cleaning methods or further modification of documents need to be changed through change control documentation (see Flow Chart 5).

Guidance to Cleaning Validation in Diagnostics

Cleaning Validation Strategy

Flow Chart 5: Monitoring Phase

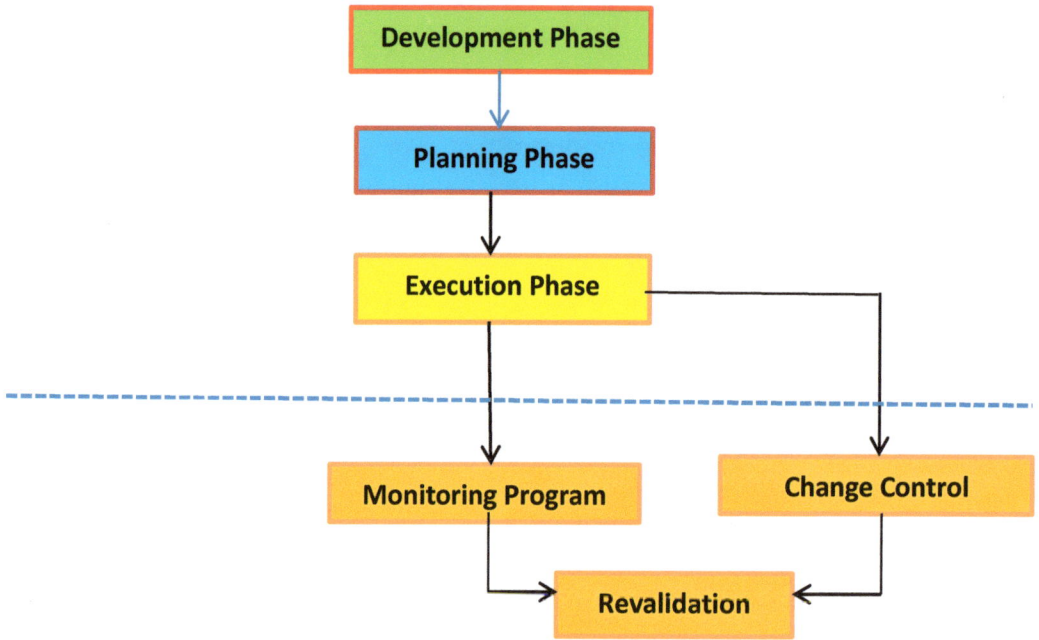

12. Verification and Validation Requirements

12.1 Analytical Procedure

The analytical procedure refers to the way the assay is performed. It should describe in detail steps that are necessary to perform each analytical test per ICH guidelines and FDA's guidance for industry for validation of analytical procedures. The test should include the testing protocol, the samples, the reference standard, preparation of reagents, types of measuring instruments used, generation of calibration curve, use of formula used in the calculation, and interpretation of results

Guidance to Cleaning Validation in Diagnostics

12.2 Specificity

Specificity is the ability to assess univocally the analyte in the presence of other components that may be expected to be present as impurities, cross reactants, degradation and unreacted products. Lack of specificity in an analytical procedure can be compensated by using defined purified reagents and/or potency of the material used.

It is not possible to demonstrate specificity (complete discrimination) using a single analytical procedure. In such cases a combination of more than analytical procedures may be necessary to achieve the level of discrimination.

12.3 Accuracy

The accuracy of an analytical procedure is defined as the closeness of agreement between the value which is accepted either as a conventional true value or an accepted reference value and the value found.

Accuracy should be assessed using a minimum of 9 determinations over a minimum of 3 concentration levels covering the specified range (e.g. 3 concentrations /3 replicates each of the total analytical procedures).

Accuracy should be reported as percent recovery by the assay of known added (spiked) amount of analyte in the sample or as the difference between the mean and the accepted true value together with the confidence intervals.

12.4 Precision

The precision of an analytical procedure is defined the closeness of agreement between a series of measurements obtained from multiple sampling of the same homogeneous sample under the prescribed conditions. Precision should be investigated using homogeneous authentic samples. However, if it is not possible to

Guidance to Cleaning Validation in Diagnostics

obtain a homogeneous sample, it may be investigated using artificially prepared samples. The precision of an analytical procedure is usually expressed as the variance, standard deviation or coefficient of variation of a series of samples.

12.5 Repeatability

Repeatability is defined as a precision under the same operating conditions over a short interval of time. Repeatability is also termed as intra-assay precision.

Repeatability should be performed using a minimum of 9 determinations covering the specified range for the procedure (e.g. 3 concentrations /3 replicates each of the total analytical procedures).

12.6 Intermediate Precision

Intermediate precision is defined within-laboratories variations: different days, different analysts, and different equipment.

Standard deviation (SD), relative standard deviation (RSD), coefficient of variation (CV) and confidence interval should be reported for reproducibility.

12.7 Reproducibility

Reproducibility is defined as the precision between laboratories (collaborative studies using the same methodology). Reproducibility should be considered in case of standardization of an analytical procedure.

Standard deviation (SD), relative standard deviation (RSD), coefficient of variation (CV) and confidence interval should be reported for reproducibility.

Guidance to Cleaning Validation in Diagnostics

12.8 Limit of Detection

Limit of Detection (LOD) of an analytical procedure is the lowest amount of analyte in a sample which can be detected. There are several approaches for determining the detection limit

a. Visual Evaluation:

It can be determined for non-instrumental (e.g. Diagnostic strip assays)

b. Signal-to-Nose Ratio:

This approach can only be applied to analytical procedures which exhibit baseline noise. Determination of signal-to-noise ratio is performed by comparing measured signals from samples with known low concentrations of analyte with those of blank samples and establishing the minimum concentration at which the analyte can be reliably detected. A signal-to-noise ratio of 2 to 1 is generally considered acceptable detection limit.

Limit of Detection (LOD)) may be expressed as: $LOD = 3.3(SD/3)$, where SD is the standard deviation and S is the slope of the calibration curve.

The standard deviation can be measured by analyzing appropriate (n=10) number of blank samples and calculating the standard deviation of these responses.

12.9 Limit of Quantitation

Limit of Quantitation (limit of an analytical procedure is the lowest amount of analyte in a sample which can be quantitatively determined with suitable precision and accuracy. Quantitation limit is a parameter of quantitative assays for the low level of compounds in sample matrices and is used for the determination of impurities and/or degradation products.

Guidance to Cleaning Validation in Diagnostics

Limit of Quantitation can be using standard deviation and slope of the response calibration curve using the equation LOQ = 10 (SD/3), where SD is the standard deviation and S is the slope of the calibration curve. The standard deviation can be measured by analyzing appropriate (n=10) number of blank samples and calculating the standard deviation of these responses.

12.10 Linearity

The linearity of an analytical procedure is it ability (within a given range) to obtain the test results which is directly proportional to the concentration (amount) of analyte in the sample.

A linear relationship should be evaluated across the range of the analytical assay procedure. It may be demonstrated by dilution of the analyte standard by dilution and running a calibration curve. If there is a linear relationship, the test results should be plotted using statistical methods by calculation of a regression line by the method of least squares. Data from the regression line should be helpful to provide mathematical estimates of the degree of linearity.

The correlation coefficient, y-intercept, slope of the regression line and the residual sum of squares should be documented. It should be noted that some analytical procedures, such as immunoassays, do not demonstrate linearity. In such cases, the analytical response curve may be non-linear and statistical tools must be used to fit the response curve (e.g. four parameter Logistic Curve, 4PLC).

For establishing linearity, a minimum of 5 concentrations is recommended (e.g. 4 concentrations /3 replicates each of the total analytical procedures).

12.11 Range

The range of an analytical procedure is the interval between the upper and lower concentration of analyte in the sample for which it has been demonstrates that the analytical procedure has a suitable level of accuracy, precision and linearity.

The range of an analytical procedure may be evaluated using samples containing analyte within or at the extremes of the specified range.

12.12 Robustness

The robustness of an analytical procedure is a measure of its capacity to remain unaffected by small, but deliberate variations in method parameters and provides an indication of its reliability during normal usage.

The evaluation of robustness of an analytical procedure should be considered during the development phase. If measurements are susceptible to variations in analytical conditions, it should be suitably controlled or a precautionary statement should be included in the procedure. During development, variations a number of system suitability test should be considered (e.g., stability of the analytical solutions, assay temperature, humidity, lot to lot variability, pH conditions, instrument to instrument variability, matrices etc.).

13. Sampling Methods

US Food and Drug Administration's "Guide to Inspection of Validation of Cleaning Process" in July 1993 prompted cleaning validations as essential in pharmaceutical and diagnostic manufacturers. Validation is required not only for manufacturing sites, but also for the sampling– filling suite in research and development. To ensure that the sampling techniques chosen meet the established acceptance criteria, pre-validation feasibility studies and method development must be performed. The two main sampling techniques available for

Guidance to Cleaning Validation in Diagnostics

cleaning validation are rinse and swab sampling. FDA prefers swab sampling to rinse sampling (1, 17, 18, 21-23).

13.1   Swab Sampling

Swab sampling is a crucial step in cleaning-sample preparation. The swabbing motion is a physical interaction between the swab and the substrate. The process generally comprises several manual steps, and requires a standardized swabbing technique to ensure that it is properly performed to have consistent recoveries from swabbing the residue from the surface. Proper training performing the swab sampling must be properly trained. Apart from this, the recovery of residual sample is dependent on the following major factors:

a. Consistent Swabbing technique

b. Operator- to- Operator variability

c. Type of protein – its solubility and stickiness to the surface

d. Material of the equipment (stainless steel, polypropylene, polycarbonate, glass, polyethylene etc.)

e. Roughness of the surface

f. Geometry of the surface

g. Drying time on the surface

h. Swab removal solvent

i. Swab extraction solvent

j. Type of assay used (Protein assay, TOC, HPLC)

A diagnostic industry manufactures a variety of in-vitro diagnostic reagents for immunoassays. A vast majority of the product formulations (calibrators, controls, enzymes, microparticles, conjugates, diluents, and wash buffers) contain plasma/serum or its protein components. These proteins are added to enhance the stability of the product and to create a native environment.

Guidance to Cleaning Validation in Diagnostics

Six types of proteinaceous materials commonly used in vitro-diagnostic product formulations reagents (abbreviated as Protein A to Protein F) are:

1. Protein A
2. Protein B
3. Protein C
4. Protein D
5. Protein E
6. Protein F

In general, these bulk proteins are used in very high concentration in comparison to an active component in any formulation. The concentration of the bulking proteins varies with each product formulation. They are in approximately 10,000 to 100,000 times higher than the active component.

The swabbing technique is intended to be used as a sampling tool for residual analytes on production equipment following cleaning. An effective strategy in challenging a cleaning method is to select a representative bulking protein (soilant) that is hardest to remove when coated and dried on equipment surface, such as, glass, stainless steel or polypropylene. It is to be expected that serum gamma globulin (SGG) being to have the lowest solubility in water would be the worst case scenario (WCS) among the five test soilants noted above and represents the hardest to remove from equipment surface.

The swab sampling procedure can also be used for non-protein components, such as, detergents, cleaning agents, lubricants, chemicals, conjugates, microparticles, and others that are used in in-vitro diagnostic reagents, provided the swab solvent, extraction solvent and the test surface is characterized. Any validated assay, such as, BCA Protein Assay or Total Organic Carbon (TOC) method can be adopted for swab recovery to support equipment validation.

Guidance to Cleaning Validation in Diagnostics

13.2 Swabbing Procedure

The procedure described here is a swab sampling procedure that can be used for swab recovery studies and also for routine swab sampling from a material surface for protein or non-protein compounds. The swab sampling procedure is intended to be used for routine verification of the effectiveness of equipment cleaning procedures and to support cleaning validation. It is recommended to use the following materials, methods and acceptable criteria.

a. **Selection of Swabs**

Texwipe TX 714A Alpha Swab is most suited for surface sampling because of minimum extractable interferences, ultra-low particles and fibers, solvent compatibility and high recovery rates. Other Swabs, such as, TexWipe 761 Alpha Swab and Puritan Dacron Swab can be used for small surface area

a. **Swab Wet Volume**

The volume of the solvent that need to be used to dampen the swab is very critical for effective swab sampling. For Texwipe 714A swabs, 200 µL of each solvent (100 µL on each side of the swab head) is recommended. For Texwipe TX 761 and Puritan Dacron swab 100 µL of total swab solvent is recommended.

b. **Swab Solvent and Swab Extraction Solvent**

Swab is a technique which allows removal of surface residue by physical entrapment and also by solubilizing the residue being swabbed. For protein swabbing, 0.1% SDS in water is recommended. Low pH acidic buffer or 2-5% phosphoric acid may also be used (for TOC method). The selection of swab extraction solvent should be such that it does not interfere with the analysis of the swab extracts and analyte being swabbed remain in solution. It is always preferred that to

Guidance to Cleaning Validation in Diagnostics

use the same extraction solvent as that of swab solvent. For protein swabbing 0.1% SDS is recommended for both swab solvent and swab extraction solvent.

c. Swab Sampling

For equipment surface sampling, the surface area equivalent to about 5 cm X 5 cm need to be identified. Swab sampling of the designated area should be done with a moistened swab. Twelve (12) horizontal strokes in one direction with side #1 of the swab head followed by twelve (12) vertical strokes of the same designated area in one direction with side#2 of the swab head provide the best results. This technique is schematically shown in Figure 1.

Guidance to Cleaning Validation in Diagnostics

Step 1: Horizontal Swabbing Technique

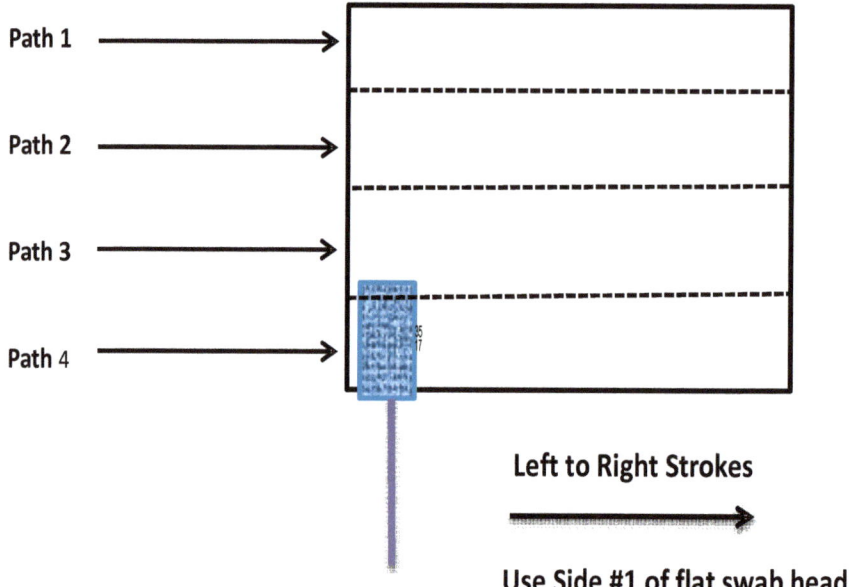

Path 1

Path 2

Path 3

Path 4

Left to Right Strokes

Use Side #1 of flat swab head

Step 2: Vertical Swabbing Technique

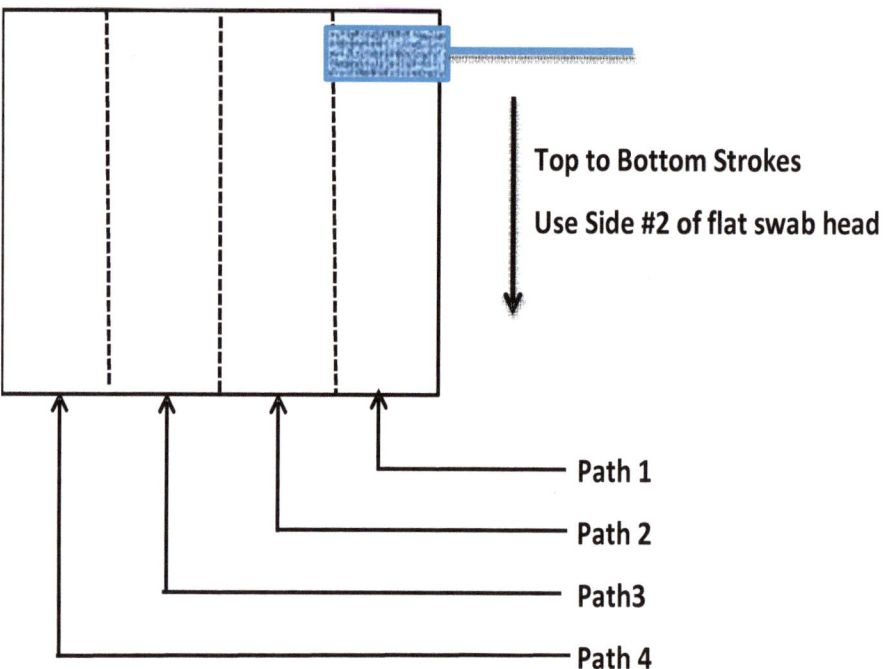

Top to Bottom Strokes

Use Side #2 of flat swab head

Path 1

Path 2

Path3

Path 4

Figure 1. Swabbing Procedure for Removal of Surface Residues

Guidance to Cleaning Validation in Diagnostics

13.3 Swabbing Recovery of Proteins by Protein Assay

d. Coating of Proteins on Material Surface

Swab recovery of protein is dependent on the protein itself and also on the deposition method. Higher recovery is obtained by spreading protein solution uniformly rather depositing in high concentration in a confined area. Glass and stainless steel can be uniformly by spraying or spreading. However, hydrophobic surfaces such as polypropylene, polyethylene or polycarbonate need to be coated as micro dots over the designated area to be swabbed. The protein recovery is also dependent on the time the protein is dried on the surface (hold time).

The swab recovery of swabbed protein sample is calculated using the following formula:

% Protein Recovery=

(Conc of Protein Swabbed – Ave Swab Blank (µg/mL) X Vol of Swab Soln (Amount of Protein Spiked (µg) X100)

e. Selection of Protein Assay Method in Swab Recovery

Pierce's Coomassie Blue Plus and Pierce's Bicinchoninic Acid (BCA) micro protein assays are generally used for swab recovery of proteins. However, it is known that higher recovery of protein is obtained when a detergent SDS is used. It is known that Coomassie Blue protein assay is not compatible with detergents and the color response is highly dependent on the time the absorbance is read, it is better to use BCA micro protein assay due to its non-interference with most of the detergents and highly reproducibility. For this reason, the use of BCA is recommended for the swab recovery of proteins.

Guidance to Cleaning Validation in Diagnostics

f. Swab Recovery with BCA Protein Assay

Swab recovery with optimized swabbing technique as described above using BCA micro protein assay showed consistent high recovery of Protein A on glass surface. The average recovery is between 91% and 94% when 15 µg to 25 µg of Protein A is coated on a 26 sq.cm area and dried for 24 hours.

g. Swab Recovery of Proteins on Different Material Surfaces

Swab recovery of other proteins including the worst case scenario (WCS) Protein B, Protein A, Protein B, Protein C, Protein D, Protein and Protein F coated on glass, stainless steel, polypropylene and polycarbonate was evaluated with the optimized swabbing technique and micro protein BCA protein assay. The average recovery of proteins on glass, polypropylene and polycarbonate was between 90% and 95%. However, the recovery from stainless steel was lower and was between 68% and 81%.

The recovery results should be considered excellent because it is not possible because of the intrinsic solubility of the residue and the nature of the manufacturing surfaces; however, it is generally not possible to achieve recovery beyond a certain level (17, 18, and 21). FDA guidelines recommend a minimum of 50% recovery (1, 21).

h. Swab Recovery Dried on Different Materials for Different Times

Swab recovery of WCS Protein B coated on glass, stainless steel, polypropylene and polycarbonate surfaces for different times with the optimized swabbing technique shows high dependence of the coated surface material and drying time. The average recovery of WCS Protein B in glass, polypropylene and polycarbonate is between 84% and 97% up to 72 hours. However, the recovery of WCS Protein B on stainless steel is very much dependent on the drying time. The recovery is about 74% in 4 hours, 40% in 24 hours and 25% in 72 hours of drying.

14. Total Organic Carbon (TOC) Method

A TOC analyzer (e.g., TOC Shimadzu 5000A or others) can be used to measure carbon in liquid samples. The following principles are employed: Total Carbon (TC) is measured by combustion of any samples that contain carbon. The carbon containing test sample is injected into the combustion tube and the carbon is catalytically converted to CO_2 at $680^{\circ}C$. Carrier gas, ultrapure air, flowing into the combustion tube transports the CO_2 of the sample through the reaction vessel to the analyzer. Excess water is removed from the sample gas. The carrier gas stream containing the CO_2 then travels through a halogen scrubber into the sample cell of the Non-Dispersive Infrared (NDIR) detector where the CO_2 is detected. The NDIR detector outputs an analog signal that generates a peak whose area is calculated by a data processor. The peak area is proportional to the TC concentration of the sample.

Organic Carbon, in the form of Non-Purge able Organic Carbon (NPOC) is can also be similarly determined. Samples are first acidified to convert all inorganic carbon species to CO_2. The carbon dioxide (purgeable) is removed by sparging and bubbling of ultrapure air through the sample. The treated sample is then injected into the combustion tube. Non-purgeable Organic Carbon is oxidized to CO_2 in this step and analyzed as Total Carbon. The instrument is standardized prior to sample testing by calibrating the instrument with standards with known carbon content. The instrument prints out Mean Area Counts (MAC) and the concentration of carbon in units of ppb (parts per billion), after replicate injections of the same sample.

The TOC method using Shimadzu TOC-5000A Analyzer is capable of determining from 15 ppb (ng/mL) to 10,000 ppb (μg/mL). For protein assay in swab sampling the recommended range is between 300 ppb and 1000 ppb. For protein recovery by the optimized swabbing technique (as described above), the same procedure using 1% SDS as wetting solvent is used followed by cutting the swab head and dropping into a cleaned Shimadzu sample vial containing 40 mL of 5 mM HCl.

14.1 Swab Recovery with TOC Method

It is important to note that elemental carbon analysis must be known before any TOC analysis which is necessary for the calculation of the concentration of samples from ppb results obtained from TOC analysis. Swab recovery of protein is dependent on the protein itself and also on the deposition method. Higher recovery is obtained by spreading protein solution uniformly rather depositing in high concentration in a confined area. Glass and stainless steel can be uniformly by spraying or spreading. However, hydrophobic surfaces such as polypropylene, polyethylene or polycarbonate need to be coated as micro dots over the designated area to be swabbed. The protein recovery is also dependent on the time the protein is dried on the surface (hold time).

The swab recovery of swabbed protein sample is calculated using the following formula:

% Carbon Recovery=

(ppb Swabbed from Surface – ppb Swab from Surface Blank X100)
ppb Coating Solution – ppb coating Solution Blank)

Swab recovery with optimized swabbing technique as described above using TOC method showed consistent high recovery of Protein A on glass and polypropylene surface. The average recovery TOC method is between 65% and 90% when 80 μg to 160 μg of Protein A is coated on a 26 sq.cm area and dried for 24 hours. The % recovery of BSA is between 30% and 50% with 40 μg Protein A.

Guidance to Cleaning Validation in Diagnostics

i. Swab Recovery of Other Proteins on Different Material Surfaces by TOC

Swab recovery of other proteins including the WCS Protein B, Protein A, Protein C, Protein D, Protein E and Protein F coated on glass, stainless steel, polypropylene and polycarbonate can be evaluated with the optimized swabbing technique and TOC method. The average recovery of all proteins on glass, polypropylene and polycarbonate was between 70% and 90%, both by BCA micro protein and TOC assays. However, the recovery from stainless steel is much lower and very much dependent on the amount of protein coated and drying time. The % recovery of all proteins including WCS Protein A on stainless steel runs between 30% and 60% when coated with 80 µg to 160 µg of WCS Protein B on a 26 sq.cm area and dried for 24 hours. It is to be noted that the % recovery with much smaller content (10 µg to 25 µg samples) would provide much higher % recovery with BCA protein assay with good precision for all proteins including the WCS Protein B and is highly recommended. However, it should be noted that the recovery of protein by Swab Recovery is not quantitative and it varies from protein to protein and also the surface of the equipment used. So, it is important to determine the Swab Recovery Factor (SRF) values of the proteins that are being used in the preparation of the diagnostic reagents. So For use of the % recovery with TOC assay, proper corrections are made using a Swab Recovery Factor (SRF) (% Recovery/100) in calculation of results. It should be noted that the SRF value for that particular equipment surface must be used in the actual % recovery calculation.

Guidance to Cleaning Validation in Diagnostics

Following is an example how to calculate % recovery of WCS Protein B in ppb from Swab Recovery (parts per billion) (% of Carbon in WCS Protein B is 49.01)

$$\% \text{ Recovery by TOC} = \frac{45 \,\mu g \text{ WCS Protein B } (49.01 \text{ g C})(1000 \text{ ng})}{40 \text{ mL } (100.0 \text{ g WCS Protein B}) (1 \,\mu g)}$$

$$= 551 \text{ ppb Carbon}$$

$$\text{Actual \% Recovery by TOC} =$$

$$\frac{45 \,\mu g \text{ WCS Protein B } (49.01 \text{ g C}(1000 \text{ ng}) \text{ X SRF}}{40 \text{ mL } (100.0 \text{ g WCS Protein B}) (1 \,\mu g)}$$

$$= 551 \text{ ppb Carbon X SRF}$$

14.2   Rinse Recovery Procedure

Rinse recovery is another sampling procedure that allows sampling of a large surface area. In addition, an inaccessible area of equipment that cannot be routinely disassembled is evaluated by this procedure. A disadvantage of rinse samples is that the residue or contaminant may not be soluble or may be physically occluded in the equipment. *FDA recommends "checking to see that a direct measurement of the residue or contaminant has been made for the rinse water when it is used to validate the cleaning process".*

To evaluate the effectiveness of rinse recovery, four major bulking proteins Protein A, WCS Protein B, Protein C and Protein D that are used in most of the calibrators, controls and/or buffers need to be tested. Cleaned stainless steel, polypropylene, and glass beakers are used. For stainless steel and glass beakers coating with the protein solutions are done with small volume (0.35 μL) of 0 μg to 25 μg protein concentrations and for polypropylene the protein solutions are added to the bottom of the beakers as micro dots. All beakers are left overnight at room temperature to evaporate to leave a thin

Guidance to Cleaning Validation in Diagnostics

film at the bottom of the beakers. The beakers are then rinsed with 3 mL of distilled water by mixing on an Orbital Shaker for 5 minutes. The rinsed water from each beaker is transferred to glass scintillation vial and aliquots of each are analyzed by BCA micro protein and TOC assays after proper dilutions.

For data analysis, the surface area of the bottom of each beaker (approximately 25 cm^2) is calculated according the geometric formula:

Surface Area = πr^2

The percent recovery of each protein is calculated by dividing the amount (µg) of protein in the rinse water from the dried beakers by the following equation:

% Recovery = $\dfrac{100 \times (\text{µg Protein in Rinse Water})}{(\text{µg Protein in Coating Solution})}$

It is to be expected the % recovery in the rinse samples to be poor due to the fact these proteins stick to the surface, thereby yielding poor results due to their solubility in water. Therefore, a Rinse Recovery Factor (RRF) must be used to calculate the actual % recovery. It varies for each protein. As an example, the RRF of Protein A and Protein C on polypropylene is 1.0 because the % recovery by rinse samples is nearly quantitative (>95%) and on glass the RRF of Protein A is 0.70 and Protein C is 0.40. The RRF values on stainless steel also follow similar trend with RRF of 0.20 for both Protein A and WCS Protein B. So, it is important to determine the RRF values of the proteins that are being used in the preparation of the diagnostic reagents. In order to use Rinse Recovery, a Rinse Recovery Factor (RRF) must be used for the type of equipment cleaned. The following equation is used to calculate the actual % Recovery

Actual % Recovery

= $\dfrac{100 \times (\text{µg Protein in Rinse Water}) \times \text{Rinse Recovery Factor}}{\text{g Protein in Coating Solution})}$

Guidance to Cleaning Validation in Diagnostics

15. Chlorine Test

Bleach Solution (Sodium Hypochlorite) is commonly used as a sanitizing and/or cleaning agent in in-vitro diagnostic industries. The presence of sodium hypochlorite over certain concentrations is known to cause deactivation of products and also it interferes with the BCA micro Protein assay. It is therefore essential to remove sodium hypochlorite contamination from equipment surface that will be in direct contact with in-vitro diagnostic products.

The Chlorine Test is available as a kit from Chemetrics. It employs a proprietary compound DDDP (N, N-diethyl-N, N -diethyl-p-phenylenediamine) and is used to for chlorine content in water using a Color Comparator. DDDP is a derivative of DDP (N, N-diethyl-p-phenylenediamine) and is oxidized by free chlorine to form a purple colored species. When ammonia or amines are present in the sample, chlorine may exist as "combined chlorine". Total chlorine is a combination of free chlorine and combined chlorine. Total chlorine (sum of free and combined) concentration is determined by adding an Activator Solution A-2500 (potassium iodide), also supplied by Chemetrics. The concentration of the total chlorine is evaluated by reading with the Color Comparator within one minute. The intensity is directly proportional to the total chlorine concentration.

The Chemetrics Color Comparator method is not very accurate and is highly subjective due to operator-to-operator reading variability. A method for accurate and quantitative method can be developed using Chemetrics reagents. Standard chlorine solutions are prepared from sodium hypochlorite (Sunbrite Bleach) from 0.10 ppm to 5.0 ppm and assay using the Chemetrics reagent. Instead of using Color Comparator, the solutions are read on a UV-Visible spectrophotometer at wavelength 562 nm or 612 nm (both have similar extinction coefficients) to generate a calibration curve. The calibration curve is linear over a range of 0.10 ppm to 1.0 ppm and 1.0 ppm to 5.0 ppm. A quantitative method for chlorine can be easily developed using standard analytical development procedure as described 14.1. Any unknown sample can be quantitated from the regression equation of the calibration curves. This chlorine test can be suitably used to detect presence of hypochlorite on any equipment surface (Chemetrics, 4295 Catlett Road, Midland, VA 22728, U.S.A).

16. Phosphate Test

In natural water and wastewaters, phosphate (P) exists solely as phosphates. Phosphates exists in three forms: orthophosphates, condensed phosphates (pyro-, meta- and other polyphosphates), and organically bound phosphates. Here a method for quantitation of orthophosphates is described.

In this test method, a mixture of ammonium molybdate and potassium antimonyl tartrate is used which react with orthophosphates to form an antimonyl-phosphate-molybdate complex. The complex on reduction with L-ascorbic acid forms a deep-colored Molybdenum Complex. The color intensity is directly proportional to the phosphorous concentration and a method for accurate and quantitative method is developed using the ammonium molybdate/antimonyl tartrate/L-ascorbic acid method. Standard phosphate standards solutions are prepared using potassium dihydrogen phosphate (KH_2PO_4) from 0.10 ppm to 3.0 ppm. The reaction mixture with the ammonium molybdate/antimonyl tartrate/L-ascorbic acid is then read on a UV-Visible spectrophotometer at wavelength 715 nm to generate a calibration curve. The calibration curve is linear over a range of 0.10 ppm to 3.0 ppm. A quantitative method for phosphate can be easily developed using standard analytical development procedure as described 14.1. Any unknown sample can be quantitated from the regression equation of the calibration curves.

This phosphate test method can be suitably used to quantitate orthophosphate concentration in rinse samples that uses orthophosphate containing detergents, such as, Mandate (see Table 1). The detergents Advance and Principal (see Table 1) contain polyphosphates and they need to be hydrolyzed by gentle digestion with sulfuric acid to orthophosphates prior to the test. The converted orthophosphates reaction mixture can be tested with the ammonium molybdate/antimonyl/ tartrate/L-ascorbic acid method.

Guidance to Cleaning Validation in Diagnostics

17. Conductivity Test

Conductivity test is used as an indicator of the removal of ionic residue. To use conductivity as a validation of cleaning, the acceptance criterion is that the rinse water is below the Limit of Quantitation Limit (LOQ) of the incoming water for the cleaning agents used. For measurement of conductivity, the conductance meter must be calibrated and standardized. For determining the efficacy of cleaning, pure distilled water can be spiked with known amounts expected impurities. In this case, cleaning agents (see Table 1) can be chosen as expected impurities. Calibration standards (100, 50 and 10 micromhos/cm (VWR Scientific, USA) are first read with the conductance meter, followed by conductivity measurements of different concentrations (or dilutions) of detergents. The measurements are done by dunking the conductivity cell few times in the liquid sample in order to flush the electrodes and remove air bubbles.

The results are then plotted with conductivity values in y-axis and concentrations in x-axis and linear relationships are observed for the detergents listed in Table 1. Regression equations are calculated by the method of least squares with r^2 >0.995 for all detergents tested. Limit of Quantitation (LOQ), the lowest amount of a detergent in a sample that can be determined with accuracy and precision is calculated from LOQ = 10(SD/slope). The LOQ values are different from each other, but once established, can be used as an acceptable limit for equipment cleaning in the rinse samples.

18. Acceptable Limits

It is a difficult task to set up an "acceptable limit" for in-vitro diagnostic products. This is the statement from FDA on this subject mostly based on pharmaceutical and biotechnology industries. The statement says the following:

"FDA does not intend to set acceptance specifications or methods for determining whether a cleaning process is validated. It is impractical for FDA to do so due to the wide variation in equipment and products used throughout the bulk and

Guidance to Cleaning Validation in Diagnostics

finished dosage form industries. The firm's rationale for the residue limits established should be logical based on the manufacturer's knowledge of the materials involved and be practical, achievable, and verifiable. It is important to define the sensitivity of the analytical methods in order to set reasonable limits. Some limits that have been mentioned by industry representatives in the literature or in presentations include analytical detection levels such as 10 PPM, biological activity levels such as 1/1000 of the normal therapeutic dose, and organoleptic levels such as no visible residue. In establishing residual limits, it may not be adequate to focus only on the principal reactant since other chemical variations may be more difficult to remove".

To determine an acceptable limit for in-vitro diagnostic products, the strategy that seems most rationale is to determine the maximum carryover concentration limit for an active analyte. The most effective way to establish the majority of products, the first thing is to select a worst case scenario analyte from different categories from the list of products, reagents and chemicals that are used in manufacturing.

The following categories seem logical in many in-vitro diagnostic industries.

1. Analytes that have very high assay sensitivity, such as,

    a. Hormones, such as, human Thyroid Stimulation Hormone, human Chorionic Gonadotropin, Brain Natriuretic Peptide, Progesterone, Estradiol, Erythropoietin

    b. Hepatitis, such as, Hepatitis Surface Antigen, Hepatitis C, HIV, Cardiovascular, such as Cardiac Troponin I, Cardiac Troponin I, BNP

    c. Cancer, such as, CA 19-9, CA 15-3, Calcitonin, Prostate Specific Antigen, human Interleukin IL-2, IL-6, IL-17

2. Reagents that very sticky, such as, drug metabolites like Cannabinoids, Surfactants like Polymeric Merquat and Chemical Additives like Sodium Dodecyl Sulfate (SDS), Anti-foaming agents

Guidance to Cleaning Validation in Diagnostics

3. Compounds that are very hydrophobic with aqueous solubility that possess an equipment cleaning risk, such as, drugs like Carbamazepine and Gentamycin, Polyamines, Steroids

4. Analyte candidates that has potential for carryover to assay interference, e.g. Antibody-conjugates with high acridinium content, Antibody candidates with high peroxidase content (poly-HRP), alkaline phosphatase

5. High percentage of solid that can alter the physical property that may hinder cleaning effectiveness, e.g., Antibody-coated microparticles,

The objective of the study is to determine the worst case Critical Concentration Limit (CCL) for the selected number of specific active analyte reagents in liquid state and to determine which analyte represent the lowest CCL. This value may be used to establish maximum allowable Residual Cleaning Limit (RCL) that can be used in the cleanliness of manufacturing equipment. An acceptable RCL will depend on the batch volume and the surface area of the equipment in which the product is manufactured. It will also depend on the type of sampling and test methods used in the analysis and the recovery of the residue from the equipment surface.

18.1 Critical Concentration Limit

18.1.1 Immunoassay

To determine the Critical Concentration Limit (CCL) is to determine which analyte possess the maximum risk of carryover contamination. In general, in immunological reagents, there could be very high number of reagents and chemicals. To make it reasonable, five different categories of challenging analytes and associated chemicals are selected as described above.

Critical Concentration Limit (CCL) is determined by titrating a High concentration Positive Analyte to the Negative Sample to

Guidance to Cleaning Validation in Diagnostics

the point at which a false positive signal occurs. CCL is defined as the concentration at which the negative signal becomes 2SD above the Limit of Detection (LOD which defined as 2SD of the negative sample).

In general, the procedure is to select Immunoassay Calibrators of an assay and follow the procedure as follows:

a. Run the all Calibrators (usually 6) in replicates (n=3)

b. Run replicates of 10 with the Negative Calibrator to Mean $\pm$ 2SD (LOD)

c. Titrate Highest Concentration Calibrator to the Negative Calibrator

d. Determine the CCL as the concentration at which the Negative Calibrator signal becomes 2SD above the Limit of Detection (LOD)

e. Plot a linear regression plot with regression equation and r^2 (experimental Analyte Concentration as Y axis vs. Analyte Dilution µL/mL as X axis

f. If the intercept is higher than 2SD value of Negative Calibrator, force the intercept through zero intercept and plot forced through Zero intercept to get the low end of the dynamic range.

Example Calculation:

a. Calculate 2SD of the Lowest Positive Calibrator, e.g. 2SD = 0.0910 µg/mL)

b. Calculate the CCL using the linear regression equation:
   High Calibrator = 0.0910/0.0196 = 4.6 µL/mL is the CCL

c. The experimentally determined CCL value is 4.6 µL of the Highest Calibrator in Negative Calibrator

The results from all five (5) categories listed above showed ultrasensitive compound hHX has the lowest CCL of 0.85 µL. This means that 0.85 µL of the Highest Concentration Calibrator per mL of the Negative Calibrator can be tolerated without producing false positive result. This value is the same as mean value plus 2SD which is defined as Limit of Detection of the assay.

18.1.2 Rinse Recovery of Analyte/Chemicals

To evaluate the worst case scenario analytes as described above under Immunoassay, rinse recovery is determined by coating the analytes at the bottom of glass, polypropylene and stainless steel beakers and analyzing by BCA Protein and TOC assays as described above under Rinse Recovery of Proteins.

By immunoassay method, an average recovery of 20% is observed with the worst case analyte hHX compared to other analytes and chemicals tested which show average 60% to 100%. The % recovery by BCA Protein and TOC assays were 100% for hHX and for others an average recovery by both assays range from 80% to 100%.

18.1.3 Critical Concentration Limit Calculation

Example Calculation:

Uncorrected hXH CCL

a. Found CCL from immunoassay titration is 0.85 µL of Highest Concentration Calibrator in Negative Calibrator = 0.00085 mL

b. From TOC analysis the found Carbon of Highest Calibrator in Negative Calibrator =15,000,000 ppb/mL

c. Therefore, CCL of TOC for 0.00085 mL

= 0.00085 mL X 15,000,000 ppb/mL = 12,750 ppb Carbon/mL in Negative Calibrator

Guidance to Cleaning Validation in Diagnostics

<u>Corrected hHX CCL</u>

a. To get the correct CCL, elemental analysis for Carbon is evaluated. As the majority of the bulking protein is Protein A, it can be used in this correction. Elemental analysis shows that Protein A contains 50.57% Carbon

b. Run a TOC assay with known concentrations of Protein A in ppb and measure the TOC values in ppb

c. Plot a linear regression curve with regression equation and r^2 with known concentrations as X axis and TOC found values as Y axis

d. Calculate the corrected CCL Carbon for hHX in ppb using the equation

Y (Found Carbon) = 0.883X (Actual Carbon) – 59.445 (Intercept)

$$\text{Corrected TOC Carbon} = \frac{(12,750 \text{ ppb} + 59.445 \text{ ppb})}{0.883}$$

= 14,500 ppb Carbon   = 14.5 ppm

18.1.4   Conversion of CCL Carbon to Total Protein

a. Majority of the bulk protein in the Calibrators contain Protein A

b. Carbon content in Protein A = 50.57% (elemental analysis)

c. Conversion Factor = 100/50.57 = 1.977

d. Therefore, Protein CCL = 14.5 X 1.977 = 28.7 ppm

Guidance to Cleaning Validation in Diagnostics

18.2   Residual Cleaning Limit Calculation

Residual Cleaning Limit (RCL) is the specification limit that defines the PASS/FAIL status in cleaning equipment. With the elaborate studies in common diagnostic assays, it is necessary to determine which ultrasensitive analyte possesses the maximum risk of contamination in cleaning processes. In this book an example has been provided to determine how to determine RCL. This must be determined for a diagnostic manufacturer to establish on the basis of the reagents and proteins used in its products. Here we have taken hHX represents the worst case scenario in terms of possible carryover contamination in manufacturing equipment. On this basis, the calculated limit is 14.5 ppm Carbon (as measured TOC) or 28.7 ppm total protein (see discussion above). Total Protein can be suitably used as an "Acceptable Limit" for PASS/FAIL in the cleanliness of production equipment for in-vitro diagnostic manufacturers.

The following are few examples of RCL calculation using BCA Protein and TOC assays:

Guidance to Cleaning Validation in Diagnostics

18.2.1 RCL Calculation using Total Protein and Swab Sampling

$$RCL = \frac{CCL \times V \times SSA \times SRF}{A}$$

Where,

RCL = Residual Cleaning Limit in units of µg/mL (ppm) /Swab. This is the specification limit that defines the PASS/FAIL criteria of the Swab/Protein

CCL = Critical Concentration Limit in units of µg/mL (ppm)(Corrected)

V = Minimum Batch Volume in mL for an equipment in which the product will be manufactured

SSA = Swab Surface Area in square centimeters (usually 5 cm X 5 cm)

A = Maximum Amount of Surface Area in units of square centimeters that will be in product contact

SRF = Swab Recovery Factor

Guidance to Cleaning Validation in Diagnostics

18.2.2 Example for Total Protein and Swab Sampling

The following is an example to demonstrate the RCL calculation for a 1 Liter batch of a Worst Scenario Calibrator in a 10 liter glass carboy which has the product contact area of 2077 cm^2. (CCL Protein = 28.7 µg/mL (ppm)

RCL of Protein in µg/mL (ppm)

$$= \frac{CCL\ Protein\ in\ \mu g/mL\ (ppm)\ X\ V\ X\ SSA\ X\ SRF}{A}$$

$$= \frac{28.7\ in\ \mu g/mL\ (ppm)\ X\ 1000\ mL\ X\ 25\ cm^2\ X\ SRF}{2077\ cm^2}$$

$$= 345.5\ \mu g/mL\ (ppm)\ X\ SF$$

In BCA Protein Assay the volume used is 3 mL, then

RCL Protein = 345 µg/mL (ppm)/3 mL = 115 µg/mL (ppm) X SRF

Assuming Swab Recovery Recover is 80%, the SRF = 0.80

Therefore, RCL of the Protein in µg/mL (ppm)

$$= 115\ \mu g/mL\ (ppm)\ X\ 0.80 = 92\ \mu g/mL\ (ppm)$$

Guidance to Cleaning Validation in Diagnostics

18.2.3   RCL Calculation using Total Organic Carbon and Swab Sampling

$$RCL = \frac{CCL \times V \times SSA \times SRF}{A}$$

Where,

RCL = Residual Cleaning Limit in units of ng/mL (ppb) /Swab. This is the specification limit that defines the PASS/FAIL criteria of the Swab/Protein

CCL = Total Organic Carbon Limit in units of ng/mL (ppb)(Corrected)

V= Minimum Batch Volume in mL for an equipment in which the product will be manufactured

SSA = Swab Surface Area in square centimeters (usually 5 cm X 5 cm)

A = Maximum Amount of Surface Area in units of square centimeters that will be in product contact

SRF = Swab Recovery Factor

Guidance to Cleaning Validation in Diagnostics

18.2.4 Example for Total Organic Protein and Swab Sampling

The following is an example to demonstrate the RCL calculation for a 10 mL batch of a Worst Case Scenario Calibrator in a 10 mL glass graduating cylinder which has the product contact area of 29.8 cm^2. (CCL TOC Carbon = 14,500 ppb =14.5 ppm)

RCL of Protein in μg/mL (ppm)

$$= \frac{CCL\ Carbon\ in\ ng/mL\ (ppb)\ X\ V\ X\ SSA\ X\ SRF}{A}$$

$$= \frac{14,500\ (ppb)\ X\ 10\ mL\ X\ 25\ cm^2\ X\ SF}{29.8\ cm^2}$$

= 121,600 ppb Carbon

If TOC analysis is done 40 mL Volume/Swab, then

RCL Carbon = 121,600 ppb Carbon)/40 mL = 3,040

If TOC analysis is done 25 mL Volume/Swab, then

RCL Carbon = 121,600 ppb Carbon)/25 mL = 4,860 ppb X SRF

Assuming Swab Recovery is 80%, the SF = 0.80

Therefore, RCL Carbon in 40 mL = 3,040 X 0.80 = 2,430 ppb/Swab

RCL Carbon in 25 mL = 4,860 X 0.80 = 3,890 ppb/Swab

Guidance to Cleaning Validation in Diagnostics

18.2.5   RCL Calculation using Total Protein and Rinse Sampling

$$RCL = \frac{CCL \times V \times RSA \times RRF}{A \times RV}$$

Where,

RCL = Residual Cleaning Limit in units of µg/mL (ppm) /Rinse Sample. This is the specification limit that defines the PASS/FAIL criteria of the Rinse Protein

CCL = Critical Concentration Limit in units of µg/mL (ppm) (Corrected)

V = Minimum Batch Volume in mL for an equipment in which the product will be manufactured

RSA = Rinse Wash Surface Area in square centimeters (5 cm X 5 cm)

A = Maximum Amount of Surface Area in units of square centimeters that will be in product contact

RV = Rinse Volume

RRF = Rinse Recovery Factor

Guidance to Cleaning Validation in Diagnostics

18.2.6 Example for Total Protein and Rinse Sampling

The following is an example to demonstrate the RCL calculation for a 10,000 mL batch of a Worst Scenario Calibrator in a stainless steel container with total surface area of 5,000 cm^{2}which has the product contact area of 1,000 cm^2. (CCL Protein = 28.7 µg/mL (ppm).

RCL of Protein in µg/mL (ppm)

$$= \frac{CCL\ Protein\ in\ µg/mL(ppm)\ X\ V\ X\ RSA\ X RRF}{A\ X\ RV}$$

$$= \frac{28.7\ in\ µg/mL\ (ppm)\ X\ 10,000\ mL\ X\ 1,000\ cm^2\ X\ RRF}{1,000\ mL\ X\ 5,000\ cm^2}$$

$= 57.4$ µg/mL (ppm) X RRF

Assuming Rinse Recovery is 20%, the RRF = 0.20

Therefore, RCL of the Protein in µg/mL (ppm)

$= 57.4$ µg/mL (ppm) X 0.20 = 11.5 µg/mL (ppm)

Guidance to Cleaning Validation in Diagnostics

18.2.7 Example for Total Organic Protein and Rinse Sampling

The following is an example to demonstrate the RCL calculation for a 1,000 mL batch of a Worst Case Scenario Calibrator in a 10,000 mL polypropylene carboy which has the product contact area of 5,000 cm^2. (CCL TOC Carbon = 14,500 ppb = 14.5 ppm)

RCL of Protein in µg/mL (ppm)

$$= \frac{CCL\ Carbon\ in\ ng/mL\ (ppb)\ X\ V\ X\ RSA\ X\ RRF}{A\ X\ RV}$$

$$= \frac{14,500\ (ppb)\ X\ 10,000\ mL\ X\ 1,000\ cm^2\ X\ RRF}{1,000\ mL\ X\ 5,000\ cm^2}$$

$$= \textit{29,000 ppb Carbon X RRF}$$

Assuming Rinse Recovery is 50%, the RRF = 0.50

RCL Carbon in Rinse Sample = 14,500 ppb/Rinse Sample

Guidance to Cleaning Validation in Diagnostics

19. Conclusion

The cleaning validation program should be based on the following major elements:

a. Appropriate cleaning agents and methods

b. Detailed cleaning procedures

c. Established standard operating procedure (SOPs)

d. Good training program

e. Characterization study to determine what degree of evaluation is required to validate the procedure

f. Appropriate validation protocols to meet the challenge of cleaning SOPs

g. Acceptance criteria for Validation

h. Perform validation studies in accordance to the protocol

i. Develop sampling methods (Swab and Rinse)

j. Perform Swab and Rinse recovery studies with hard to clean reagents including cleaning agents, chemicals and/or additives

k. Identify the worst case scenario (WCS) reagent(s) that are hard to clean and establish recovery factor(s)

l. Develop test methods (TOC, Protein, Phosphate, Chlorine, Conductivity) and validate per written protocol

m. Compile, document and approve validation report for test methods

n. Execute validation with equipment cleaning procedure(s) using the acceptance criteria that includes safety factors, such as, Swab and Rinse recovery

o. Compile, document and approve validation report for equipment cleaning procedure(s)

p. Identify reagents and processes used in manufacture of in-vitro product and select worst case scenario (WCS)

q. Establish critical concentration limit (CCL) from worst case scenario (WCS) active analyte(s) that includes safety factors, such as, Swab and Rinse recovery

r. Establish residual concentration limit (RCL) using the established CCL value with protein assay and/or Total Organic Carbon analysis

s. Change control program in place

t. Corrective and preventive (CAPA) for any non-conformances in place

u. Revalidation policy in place

v. Monitoring program in place

w. Periodic auditing and reviews in place to ensure compliance

20. Acknowledgments

I am highly thankful to Abbott Laboratories for using to illustrate few points described in this book. My sincere thanks to Dr. Robert Marshman and Timothy Lucier for valuable input in this book

Guidance to Cleaning Validation in Diagnostics

21. References

1. FDA, "Guide to Inspection of Cleaning Validation" (July 1993)

2. FDA, "Validation of Cleaning Process" (August 1993)

3. FDA Center for Device and Radiological Health CDRH, "Quality System Inspection Technique" (1999)

4. FDA, Guidance for industry – Bio-analytical method validation (2001)

5. FDA, "Guidelines for the Validation of Analytical Methods for the Detection of Microbial Pathogens in Foods" (September 2011)

6. Guideline for Industry "Q2A Text on Validation of Analytical Procedure", ICH (March 1995)

7. Guidance for Industry "Q2B Validation of Analytical Procedures: Methodology", ICH (November 1996).

8. Guidance on aspects of cleaning validation in active pharmaceutical ingredient plants. CEFIC- APIC (December 2000)

9. S.W. Harder, "The Validation of Cleaning Procedures," Pharm. Technol. 8 (5), 29-34 (1984)

10. Fourman, G.L. and Mullen, M.V., "Determining Cleaning Validation Acceptance Limits for Pharmaceutical Manufacturing Operations," Pharm. Technol. 17(4), 54-60 (1993).Pharmaceutical Manufacturing", Pharm Technol. 15 (4), 54-60 (1993)

11. PDA Technical Report No. 29, "Points to Consider for Cleaning Validation", PDA J. Pharm.Sci. and Tech., **52** (6), 1–23 (1998)

12. "A Compendium of processes, materials, test methods and acceptance criteria for cleaning reusable medical devices" Association for the Advancement of Medical Instrumentation AAMI TIR30 (2003)

13. Walsh A. "Cleaning Validation for the 21st Century: Acceptance Limits for Active Pharmaceutical Ingredients (APIs): Part I", Pharmaceutical Engineering, 74-83 (2011)

Guidance to Cleaning Validation in Diagnostics

14. LeBlanc D A, "Establishing Scientifically Justified Acceptance Criteria for Cleaning Validation of Finished Drug Products,"Pharm.Technol. **22** (10), 136–148 (1998)

15. Ethier, J. (2005), Procedure for the validation of biological active pharmaceutical ingredients (APIs) manufacturing processes, The Official Journal of ISPE, Vol. 25 No. 2

16. Valsilevsky M A, "Cleaning Validation for Biological Products: Case Study", Pharmaceutical Engineering (November/December 1995)

17. Chudzik G M, "General Guide to Recovery Studies Using Swab Sampling Methods for Cleaning Validation," J. Validation Technol. 5 (1), 77–81 (1998)

18. Shifflet M J and Shapiro M, "Development of Analytical Methods to Accurately and Precisely Determine Residual Active Pharmaceutical Ingredients and Cleaning Agents on Pharmaceutical Surfaces," Am. Pharm. Rev., Winter, (4) 35–39 (2002).

19. Hall W E, "Your Cleaning Program: Is It Ready for the Pre-Approval Inspection" J. Validation Technol. **4** (4), 302–308 (1998)

20. Texwipe Tech Notes, Vol XI, No. 1 2.2011

21. Yang P, Burson K, Feder D, and Macdonald F, "Method Development of Swab Sampling for Cleaning Validation of a Residual Active Pharmaceutical Ingredient", Pharmaceutical Technology, 84-94 (January 2005)

22. Smith J A, "A Modified Swabbing Technique for Validation of Detergent Residues in Clean-in-Place Systems," Pharm. Technol. 16(1), 60-66 (1992)

23. Cooper, "Using Swabs in Cleaning Validation: A Review. Cleaning Validation IVT, 74-89 (1999)

24. Shimadzu Corporation, Total Organic Carbon Bulletin 5310

25. Analysis of T otal Organic Carbon, University of Massachusetts Environmental Engineering Research Laboratory, TOC SOP Ver. 3 (December 2012)

Guidance to Cleaning Validation in Diagnostics

26. CDC SWS Project, Chlorine Residual Testing Fact Sheet

27. WHO, How to measure residual chlorine in water, Technical Note No. 11 Draft Revised: July 2005

28. ASTM D515-88, "Standard Test Method for Phosphorous in Water"

29. Edward, Molof and Schneeman, "Determination of orthophosphate in Fresh and Saline Water", American Waterworks Association, Vol 57, p 917-926 (1965)

30. USP 25-NF 20, <1225>, "Phosphate, TOC> Validation of Compendial Methods"

31. USP 24, <645>, "Water Conductivity"

32. European Pharmacopoeia, 1997, Method 2.2.38, "Conductivity"

33. USP 24 Supplement 1, "Conductivity, Validation of Compendial Methods"

34. Alconox, "Cleaning Validation References" (2005)

www.ingramcontent.com/pod-product-compliance
Lightning Source LLC
Chambersburg PA
CBHW050756180526
45159CB00003B/1479

* 9 7 8 1 4 9 9 6 9 2 9 1 4 *